A GUIDE TO

Hazard analysis for small businesses

Donald J. MacDonald
Dagmar M. Engel

Illustrations by
Chris Woffenden

First published 1996
ISBN 1 871912 90 3

Published by
©HIGHFIELD PUBLICATIONS
"VUE POINTE", SPINNEY HILL, SPROTBROUGH,
DONCASTER, SOUTH YORKSHIRE, U.K. DN5 7LY
Telephone: Doncaster (01302) 850007
Facsimile: (01302) 311112

Printed by Garnett Dickinson Print Limited, Rotherham and London

CONTENTS

		Page
PREFACE		**1**
1	Introduction – What is HACCP?	**3**
2	HACCP – A path through the jungle of definitions	**7**
3	HACCP and the law	**15**
4	How to do it	**19**
5	Hazard analysis for small businesses	**29**
6	Last but not least – A word on documentation	**53**
APPENDICES		
I	HACCP and quality	**59**
II	The decision tree	**62**
III	Critical control points	**66**
IV	Examples of hazard analysis in small businesses	**68**
INDEX		**71**

Preface Welcome to a new system

"Illness due to contaminated food is perhaps the most widespread health problem in the contemporary world".

(FAO/WHO Expert Committee on Food Safety)

Throughout the world, food poisoning and food-borne infections are a major health problem causing many deaths and considerable illness, ranging from mild to very severe. Despite better hygiene standards and increased awareness of the causes of food poisoning, the number of reported cases continues at an unacceptable level. The actual number of cases greatly exceeds the reported number because not every affected person visits a doctor, and, even if they do, not every case is recognized or reported as food poisoning.

Analyses of food poisoning statistics show that up to 60% of outbreaks seem to be caused by poor food handling techniques and by contaminated food served in food service establishments. Apart from the obvious human effect of these illnesses they are also responsible for considerable costs to the food and health care industries and hard work for everybody involved. Their effects on the food industry are of importance to everyone from the smallest of retailers to the largest of processors. Food service establishments, no matter whether it is a hot dog stand or a high class hotel restaurant, perform a public service from which consumers have a right to expect safety as well as quality.

Several systems exist which attempt to ensure the safety of products. The traditional system is end-product testing but this has a number of disadvantages such as high costs, slow response and the involvement of external specialists. In addition, the correct interpretation of the results is highly skilled and very difficult and the method is imprecise because not all potential hazards are taken into account. (A hazard is "anything that might cause harm to the consumer".) The control is only **reactive** because remedial actions are taken **after** any calamity has taken place and even with the best statistical sampling plan there is no guarantee that a defect which exists will be found. All of this adds up to a system which may not do the job it is intended to do.

This handbook describes a control system that is **proactive,** fast, easy to monitor, cost-effective, implemented by the people directly involved with the product and, if correctly designed and executed, will achieve its objective of safe food. The practical application of this concept concerns all staff and requires a thorough knowledge of the product and the process.

The **intention** of the system is to focus control at points in the operation which are critical to the safety of the end product. The **application** of the system combines the awareness of the potential hazards associated with the food product, the process, personnel, equipment, environmental factors and the defined use of the product into an action-orientated programme which produces safer food and reduces the risk of food-borne illness. The **name** of the system is **HACCP.**

1 Introduction – What is HACCP?

HACCP – the unpronounceable word from outer space

For thousands of years people have tried to make food safer – this is amply illustrated by many of the earliest religious edicts concerning food preparation. In many ancient cultures, complex codes of practice and sacred laws concerning processing and handling of food can be found, from which have evolved modern regulations concerning the safety of food operations.

But HACCP is not a religion. It is in theory a philosophy – but in practice a tool. HACCP is the acronym for **H**azard **A**nalysis **C**ritical **C**ontrol **P**oint. It is a method of ensuring food safety by examining every step in a food operation, identifying the steps that are critical to food safety and implementing effective control and monitoring procedures at these steps.

The development of the HACCP concept was directly connected with a "food production and research" project by Pillsbury Company for space travel programmes. The fundamental idea was created through team work by Pillsbury, National Aeronautics and Space Agency (NASA), Natick Laboratories of the US Army and US Air Force Space Laboratory Project Group. It was in 1959 when Pillsbury was commissioned to produce food that could be used under anti-gravity conditions in spaceships. One problem was that nobody really knew how food or bits and fragments of food would "behave" under anti-gravity. But the main task of the project was to guarantee that the food provided for space travellers was not contaminated microbially, chemically or physically in a way that would lead a space mission either to failure or to catastrophe.

The HACCP concept was first presented in public during the National Conference on Food Protection in 1971. In 1973 Pillsbury Company published the first complete documentation on HACCP which was soon employed in the training of the staff of the US Food and Drug Administration working on acidified foods and low-acid foods. But it took another 12 years before, in 1985, HACCP was first taken into consideration for general implementation in the food industry by the National Academy of Science (NAS). Since then more and more research groups have emphasized the necessity of a preventive system to control microbial, chemical and physical hazards in food. The one-time space project became the ideal for normal daily food processing.

HACCP was originally designed to prevent food poisoning under anti-gravity conditions

In the European Union, the Council Directive on the Hygiene of Foodstuffs, 1993 sets out general hygiene principles and conditions for foodstuffs to apply throughout the food chain. Among numerous requirements it states the necessity for each food business to apply the principles of HACCP. Article 3 of the Directive requires that every activity critical to the safe production of food *must* be controlled by the food business operator.

HACCP – the philosophy

Trying to find definitions for the term "HACCP" could lead to a multi-volume dictionary. Therefore, for the moment let us concentrate on one of them – HACCP as a philosophy. It means that before making judgements about safety in general and critical points in particular, you have to have a thorough knowledge of the whole process – a way of thinking that embraces more than just food handling.

Another important aspect is that every level of staff is involved in contributing to the end product including non-technical personnel. Communication is an essential factor for the success of establishing an effective system. Everybody, not only those directly involved in the implementation of HACCP, must be made familiar with the system and feel ownership of it. A premises with a working HACCP system should operate like an ant colony – showing a never-ending flow of information and feed-back between every section including the management. The key phrase is therefore **total familiarity.**

HACCP – the tool

We have looked at the philosophical side of HACCP. Let us now concentrate on the practical side. As a powerful tool HACCP avoids the many weaknesses which are to be found in the end-product sampling, reactive approach described earlier. HACCP is a systematic approach to the identification, assessment and control of hazards, and its implementation focuses attention on the factors that *directly* affect the safety of a product at all stages of the food chain. Furthermore, the use of HACCP can lead to less wastage of product, less expensive means of control and therefore economic advantages. As a consequence, HACCP must also be applied to raw material supply as well as to the end of the chain – the final product storage and retail.

HACCP can be used to control everything – raw materials, incoming supplies, preparation, storage, processing, transportation and operations such as cleaning and disinfection. As with any tool, however, there are many ways in which to use it. In addition to being a legal, requirement it is also beneficial that every food business has a satisfactory system based on HACCP principles. It stands to reason, however, that systems may vary greatly depending on the type of business, the size of the operation and the potential food safety hazards associated with the operation, and the risks to the consumer. One of the aims of this book is to show how HACCP principles can be used even in very small businesses. As the principles used to develop HACCP work with large manufacturing companies, they should also work in the small cafe and on the ice-cream tricycle.

Even the smallest food businesses can apply the principles of HACCP

2 HACCP - A path through the jungle of definitions

After having found out that "HACCP" is **not** the Russian technical translation for "food hygiene", there is a need to explain a few of the terms that form the acronym and the ideas behind it. Soon we will find ourselves confronted with a whole jungle of definitions. But there is an easy path through this jungle, so let us start at the very beginning.

Hazard

According to the dictionary, a hazard is a danger that can occur. Staying within our picture of the jungle, there are certainly quite a lot of hazards you can experience – some of them really serious ones like wild beasts of prey, other ones not quite so threatening like insect bites. Of course, if one of those insects is a mosquito carrying malaria, an estimated "small hazard" may turn out to be a bit more serious. Drawing the conclusion that a hazard is something that might cause harm, it can be defined as follows:

"A hazard is anything – microbiological, chemical or physical – that might cause harm to the consumer."

A more complex definition of microbiological hazard is given below which, although scientifically more precise, does not roll off the tongue too easily!

"A microbiological hazard means the unacceptable contamination, unacceptable growth and/or unacceptable survival of pathogenic or spoilage microorganisms and/or the unacceptable production or persistence in foods of toxins."

Risk

Back in the jungle you may say that your greatest possible hazard may be a hungry tiger or a venomous snake or a crocodile threatening your life.

Depending on the kind of jungle you're in, the **probability** of these hazards occurring may vary greatly. For example, the risk of encountering a tiger in Malaysia is certainly much greater than the risk of becoming a tiger's lunch in the Scottish Highlands. Your personal behaviour and the protective measures you take are also of importance. The risk of being bitten by a snake will obviously be reduced by wearing a pair of strong boots. In addition, your **path** through the jungle has to be taken into account. If you walk along a river the likelihood of meeting one or two crocodiles is much greater than it would be if you took a different path far away from any water. This leads us to the term risk.

"Risk is the estimate of the probability of a hazard occurring."

In the food industry, one of the most serious **hazards,** for example, is botulism caused by the toxin of the bacterium *Clostridium botulinum* which causes serious illness and even death. The special "requirements" and the preferred sources and routes of *Clostridium botulinum* affect the **probability** of botulinum intoxication occurring. Expressed simply, this means that if the bacterium is likely to be present and conditions exist which favour its growth then there is a high risk of a problem occurring. Therefore, the risk of producing botulinum toxin is far greater in the fish canning industry than in a factory producing chocolate rumballs. It should be noted, however, that even in the chocolate rumball industry, botulinum remains one of the greatest **hazards** – although the **risk** is very small.

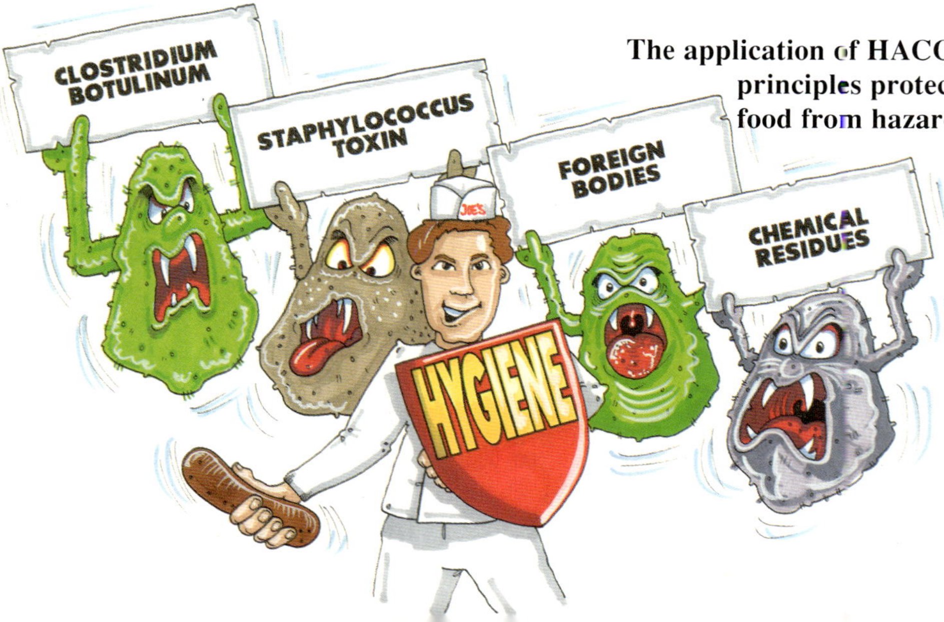

The application of HACCP principles protects food from hazards

Severity

Of course not all hazards are equally dangerous. Some hazards are life threatening whilst others cause illness or discomfort. As a consequence, we have to define the **severity** of a hazard. Going back to the jungle, you find that if a crocodile bites your leg off, it will be dangerous and extremely painful but your life, although not your limb, can be saved by quick reaction and adequate treatment. Whereas the bite of a small viper, though perhaps less painful or remarkable at first sight, leads to death almost immediately. In the same way botulism, which we have already talked about, is a very serious hazard in the food industry because it frequently leads to death whilst some other types of food poisoning, although unacceptable, merely cause extreme discomfort for a few days. We can define **severity** therefore as follows:

"Severity is the magnitude of the hazard, or the seriousness of possible consequence."

This then leads us to a definition of **hazard analysis.** Hazard analysis is:

(1) the identification of potential hazards;

(2) the risk of hazards occurring, considering:

  (a) potential sources and routes of contamination, (microbiological, chemical and physical);

  (b) the probability of survival/multiplication of microorganisms at different steps of the production; and

(3) the estimation of the severity of the hazards identified.

Risk assessment

"Risk assessment" could be defined as **"A method of estimating both the likelihood of a hazard occurring and its severity if it should occur."**

This is obviously a very complicated subject requiring expert knowledge to fully understand both the mechanisms and the way to interpret them but the following list gives an idea of some of the factors which may be taken into account when carrying out a risk assessment.

- What is the likelihood that the hazard will affect peoples' health?
- To what extent will it affect their health?
- What is the severity of the hazard?
- To what extent will public health in general benefit from controlling the hazard?
- Is the risk from the hazard imminent?

- Will the risk increase or decrease over a period of time if it is not controlled?
- How many people are likely to be exposed to the hazard?
- What is the age/health status/vulnerability of the people exposed to the hazard?
- What will each individual's exposure be to the hazard per day/event/lifetime?

From a study of these and other factors, a risk assessment can be made for any hazard in any food in any situation. The assessment is then used to help determine which hazards need to be controlled, and to what extent they need to be controlled – in **any** food operation. This kind of exercise is, however, best left to the experts.

A simpler way to look at risk assessment is to study the main causes of food poisoning. The following table, compiled using data from the USA, Canada and the UK, is a useful starting point for such an exercise.

Causes of food poisoning	
63%	Inadequate cooling and cold holding
29%	Preparing food ahead of planned service
27%	Inadequate hot holding
26%	Poor personal hygiene/infected persons
25%	Inadequate reheating
9%	Inadequate cleaning of equipment
7%	Use of left-overs
6%	Cross-contamination
5%	Inadequate prime cooking or heat processing
4%	Toxic chemicals added from containers
2%	Contaminated raw ingredients
2%	Intentional chemical additives
1%	Incidental chemical additives
1%	Unsafe sources

(Source CDC, Atlanta)

It is worth mentioning that the figures in this table do not add up to 100% because most outbreaks are caused by more than one factor.

This information can be used by even the smallest businesses to assess some of the health risks which could arise from their activities. For example, it can be seen that many of the factors on the list are concerned with poor temperature control resulting in the growth or survival of microorganisms. Therefore, when handling foods or carrying out operations where good temperature control is required, special attention should be paid to cooling times, cold holding, cooking times and temperatures, reheating, use by/best before dates and stock rotation.

This is a simple but effective method of carrying out your own risk assessment based on readily available data and common sense and which need not cost you anything other than a small investment of your time and some focused thought.

Critical control point (CCP)

Let us now leave the jungle and have a look at the second part of the acronym: CCP. We have already learnt that the three letters stand for the phrase **critical control point.** It stands to reason that every possible hazard, bearing in mind its risk of occurrence, has to be ***controlled.*** Control can be exercised at different points of the food chain depending on the kind of hazard we have to deal with. But what does "critical" mean? It may be helpful again to have a look at what the dictionary tells us:

- **of or at a crisis (when there will be a change for the better or worse)**
- **important, crucial, decisive**

(The Oxford English Dictionary; Collins Dictionary & Encyclopaedia)

We can therefore conclude that a critical control point is a step in a food operation where control is crucial to the safety of the end product (as the dictionary definition implies: a change takes place at this point and what we have to ensure is that it is a change for the better).

We can also say:

"A critical control point is a step in a system at which a loss of control would lead to an unacceptable risk to health." Or alternatively:

"A critical control point is a location, practice, procedure or process at which control can be exercised over one or more factors which, if controlled, could minimize or prevent a hazard." (ICMSF; WHO)

It is important to discuss briefly the last phrase of this definition, "minimize or prevent a hazard". In some food operations control of a CCP can *prevent* the occurrence of one or more hazards. It is, however, also possible to identify CCPs at which one or more hazard can be *minimized* but not completely prevented. In the manufacturing sector these are sometimes treated separately but the end result is still the same, they are all CCPs and therefore must be controlled.

Monitoring

Now that we have discussed CCPs it is essential to select appropriate means to check that a hazard has been controlled at a CCP. What we need are certain **criteria** that indicate whether or not the operation is under control. To ensure control of a CCP, not only do we have to select and specify appropriate criteria but also they have to be put on record in a monitoring system. What does this mean?

Monitoring is the measurement or observation at a CCP of the criteria which allows corrective actions to be taken.

Monitoring detects deviations which can then be corrected

Criteria

Methods of monitoring are numerous; most involve simple actions like visual checks, recording time and temperature and hand swabbing. But some involve complicated and ambitious procedures like complex chemical or bacteriological analysis. If we define **criteria** as characteristics which must be specified to ensure the control of product safety, the monitoring must detect any *deviation* from the specified criteria. If the monitoring system indicates that *criteria at a CCP are not met,* an appropriate remedial action has to be implemented. It is important that monitoring procedures detect the loss of control at a CCP quickly enough to take corrective action to regain control at this point and to prevent wastage of product.

Verification

Having established as much control as possible at every critical control point using monitoring, we still need to go one step further to ensure that the HACCP concept is being carried out correctly and effectively. This procedure is called **verification.**

"Verification is the use of supplementary information and tests to ensure that the HACCP system is functioning as planned."

(F Bryan)

Verification involves ***reviewing*** the entire HACCP system and its records. It can be carried out by special control staff, health personnel or by outside consultants. Verification should cover two aspects:

(1) Is the HACCP procedure as originally applied still appropriate to the product/ process hazards?

(2) Are the specified monitoring procedures and corrective actions still being properly applied?

(The HACCP Working Group, CAMPDEN, 1992)

Whereas routine monitoring of CCPs of a food operation is the responsibility of the manager, verification may involve an outside study team. Besides checking the current documentation/monitoring systems, verification should also review the composition of the product as well as any change in operational procedures that have been implemented since the establishment of the HACCP system. The results of the verification can mean that some critical control points have to be reconsidered, or that some monitoring procedures have to be modified.

We have now reached the point where we can put the components of the HACCP system together and have a look at the concept as a whole.

These bricks of the HACCP building are known as the basic principles underlying the system of HACCP. They have been formulated by the National Advisory Committee on Microbiological Criteria for Foods (1992) and the Codex Alimentarius Commission (1991). We can now clearly see that identification and monitoring of the CCPs is the heart of the system. We still do not know how to determine the CCPs of a food operation. But this will be covered in the later parts of the book. But before this let us have a close look at the legal aspects of HACCP and how HACCP interrelates with Quality Assurance.

3 HACCP and the law

We have already discussed HACCP as a system which benefits all parties: producers, processors, distributors, retailers and the consumer. It is important to emphasize that HACCP is meant to ensure food safety and that, in the European Union, conducting a food business according to “the principles of HACCP” is a legal requirement. Staying within the law is one of the basic requirements for staying in business. Furthermore, HACCP will help businesses to stay competitive and help to demonstrate due diligence.

In June 1993, the Council of the European Committee adopted the EC Directive 93/43/EEC on the hygiene of foodstuffs. This directive outlined the requirements for food hygiene, food handlers, food businesses and training of food handlers. Other product specific directives such as the directive on milk and milk products also require application of the principles of HACCP, but it is article 3 of Directive 93/43/EEC that we will use to demonstrate the EU’s approach to HACCP – as a tool to ensure food safety.

HACCP and the EC Directive

Before we plunge into the paragraphs of article 3, let us stop for a moment to have a close look at the definitions in the Directive. Some people in the food business, be they one-man-businesses or working from small premises, being producers, retailers or caterers, may have asked themselves if the European Directive affects them at all. The man working in the packaging room or the woman in charge of food storage may ask the question “Am I a food handler?” Article 2 explains who and what is covered by the Directive. There is no escape – it concerns everybody involved in the food chain.

> ***Food hygiene*** *shall mean all measures to ensure the safety and wholesomeness of foodstuffs. The measures shall cover* ***all stages*** *after primary production during* ***preparation, processing, manufacturing, packaging, storing, transportation, distribution, handling*** *and* ***offering for sale or supply*** *to the customer.*

> ***Food business*** *shall mean any undertaking, whether for profit or not and whether public or private, carrying out any or all of the following:* ***preparation, processing, manufacturing, packaging, storing, transportation, distribution, handling or offering for sale or supply of foodstuffs.***

Article 3.1 requires preparation, processing, manufacturing, etc. to be carried out "in a hygienic way". It is paragraph 3.2 that introduced an important new provision into general hygiene law in the EU – **it requires food business operators to adopt some of the principles used to develop HACCP.**

> *Food business operators shall identify any step in their activities which is* ***critical*** *to ensuring food safety and ensure that adequate safety procedures are identified, implemented, maintained and reviewed on the basis of the following principles, used to develop* ***the system of HACCP*** *(Hazard Analysis and Critical Control Point):*
>
> - ***analyzing the potential food hazards*** *in a food business operation*
> - ***identifying the points*** *in those operations where food hazards may occur*
> - *deciding which of the points identified are critical to food safety –* ***the "critical points",***
> - *identifying and implementing* ***effective control and monitoring procedures at those critical points*** *and*
> - ***reviewing*** *the analysis of food hazards, the critical control points and the control and* ***monitoring*** *procedures periodically and whenever the food business operations change.*

Although this requirement was new to general hygiene legislation, many businesses in the food manufacturing area were already familiar with HACCP. What this Directive requires is that every food business operator should, in addition to traditional principles of good premises hygiene and personal hygiene, be able to identify potential food safety hazards and control them before they result in problems. To demonstrate whether or not this provision is being observed, every food business operator needs to show that he or she has actually thought about the activities carried out in his or her business, and that effective control measures are in place for every hazard.

The Directive **does not** require the implementation of a classical HACCP system. The approach to the identification and control of hazards as well as the level of analysis and monitoring has to be modified according to the risks, size and nature of the business. Furthermore, the Directive does not require documentation but, as you will see later, it is sensible to document at least some of the procedures without the need for filing cabinets full of HACCP records. The Directive, therefore, tells you **what** to do but not **how** to do it. Later, we will discuss how you can adopt a common sense approach to HACCP and set up a system just for your business.

How HACCP, quality and the law interrelate

HACCP should be an integral part of the Quality Assurance system of a company. Or let us put it the other way round: a properly functioning QA system cannot ignore HACCP. As it is the foundation stone of the safety building we have to look at it as a compulsory factor in the QA process. But where does the legal requirement fit into the picture?

HACCP and Quality Assurance

To point out the differences as well as what HACCP, quality and law have in common, we can use an already familiar term – the critical control point. We have already learnt that defining the CCPs is the heart of the HACCP system. Whenever a CCP gets out of control, obligatory measures have to be taken to correct the production. In terms of quality, the operator of a food business is also concerned about criteria like taste, look, smell, freshness ... not "critical" control points as in HACCP, but, nevertheless, a defined control point for the quality design of the end product. Last but not least, legislation is a tool for controlling the standards of hygiene within food premises in order to prevent food poisoning. The control points are the prevention of contamination, temperature, personal hygiene, cleanliness and the provision of facilities, etc. These control points are prescribed by legislation compared to the critical status of a CCP or the defined status of a quality control point.

Let us put it all together.

	legislation	product safety	product quality
status	necessary	compulsory	voluntary
system	compliance with the law	HACCP	quality control
nature of necessary control point	prescribed	critical	defined

(Source: Corlett, 1991, modified)

Summary

- Every food business must pay attention to the control of points critical to food safety, it is a legal requirement.
- HACCP is a **food safety system** – <u>not</u> a quality system.
- A properly functioning QA system must incorporate HACCP principles.
- **You** must decide for yourself how to apply the principles of HACCP according to the risks, nature and size of **your** business.

4 How to do it

For those who are new to the concept of HACCP, the process of setting up and conducting a study looks rather complicated, it is not. It is just a sequence of simple steps whose main purpose is to ensure product safety.

In the next few pages we will look at the different steps of a HACCP study in general as quoted from the Codex Alimentarius Commission to give a glimpse into the "classical" concept. This concept is usually, and most beneficially, applied to processing operations, particularly large manufacturers. Later, we will discuss how to apply this concept to the more simple situations found in retail shops, catering premises and small processors. We will usually consider all types of hazard, microbiological, chemical and physical, unless otherwise stated in the text.

Assemble the HACCP team

Every HACCP study will require the collection, comparison and evaluation of technical data. This should be carried out by a multi-disciplinary team consisting of key persons involved in the full spectrum of the product process.

Every team member should be aware of the principles of HACCP.

Assembling a HACCP team is not the easiest thing if you are the only person in the company

The team should include representatives of all relevant staff, including technical persons, scientific specialists, management, senior executives, offering a pot-pourri of theory, experience, technical expertise and working practice. Not everyone needs to be a permanent member of the team, some may join only for those areas which concern them directly. The advantage achieved by a multi-disciplinary team is obvious, an all-over contribution to the system by everybody involved in any activity associated with the process.

It is important for all production stages that not only does the team include responsible persons dealing with a particular step of the production but also all levels of staff. This keeps the team from being too academic and results in an ideal combination of theory and practice – people describing the concept which should be put into practice and others talking about what actually happens and the practical difficulties of applying theory on the shop floor.

Describe the product

Before proceeding with any other steps it is essential to write down a full product description covering all aspects of the product. This step helps to collect information relevant to identifying hazards, assessing risks and determining control points.

Do you know all about your product?

Identify intended use

This step is designed to establish the intended use of the product by the consumer target group. It is of special importance not only to define the "normal" use by the consumer but also possible abuse of the product as far as it is within the control range of the producer.

Construct a flow diagram

A flow diagram should describe all the steps in the preparation, from raw materials through to the point of consumption. A good definition of steps is:

"not only an operation, process or procedure but also raw materials".
(Bob Mitchell)

The following list contains some of the data which may be required for the construction of a flow diagram but it is by no means exhaustive:

- raw materials/ingredients including packaging
- layout of premise/equipment
- sequence of all process steps
- time/temperature details of all raw materials, intermediate/final products
- storage and distribution conditions
- personnel routes
- personal and environmental hygiene
- cleaning and disinfection
- instructions for customer use
- potential for cross-contamination.

Again we see the importance of a multi-disciplinary team!

Construct a flow diagram

On-site verification of the flow diagram

Once we have used all this information to design the flow diagram we should ask the questions: Is the diagram realistic? – Is every step an accurate representation of the operation as it takes place daily?

This is simply an investigation of potential deviations from the original diagram. Every noted deviation can be an indication that the flow diagram has to be revised. It is also very important to verify the flow diagram during weekend and night shifts and to take account of delays in the system.

List all hazards associated with each process step and list all measures which will control the hazards.

Using the flow diagram as a skeleton we have to focus on each step and

- list all potential hazards (microbiological, physical, chemical) likely to occur at each step
- describe preventive measures which do exist or could exist to control these hazards.

"Hazards considered must be of such a nature that their elimination or reduction to acceptable levels is essential to the production of a safe food."

(Bob Mitchell)

Thinking about potential hazards involves more than trusting in what you see everyday

A determination of critical control points will not take place at this stage. At this point of the HACCP plan it is essential to project the whole spectrum of conceivable hazards without paying attention to the likelihood of their occurrence. Each identified hazard can then be confronted with one or more suitable control measures. To control a specific hazard, more than one preventive measure may be required. On the other hand, a specific preventive measure may be able to control more than one hazard, for example, cooking may destroy more than one microorganism.

Apply the decision tree to each process step in order to identify CCPs

Every HACCP model is based on the identification of CCPs. These cover microbiological, chemical and physical hazards. Steps in the flow diagram dealing with quality, good manufacturing practice and fraud will not be considered at this point. Each step in the process can be linked to one or more hazards. Applying the HACCP decision tree to each hazard will enable the team to identify the CCPs. (See Appendix II.)

When all the hazards for a particular step have been dealt with, the potential hazards at the next step in the flow diagram have to be considered. There is no limit to the number of CCPs that may be identified in the model, but it is obvious that a large number of points on which control has to be exercised affects the economy and efficiency of the process. It should also be kept in mind that control measures should be practicable, economically feasible and must ensure food safety.

Apply the decision tree to each process step

Establish target levels and tolerance for each CCP

Criteria must be specified for each step that will ensure the safety of the product.

Target level – a specified value for the control measure which has been shown to eliminate or minimize a hazard at a CCP.

Tolerance – A specified variation from the "target level" which is acceptable; values outside this tolerance indicate a deviation.

Having identified all CCPs, the target levels and tolerances for each control point have to be determined. Specified target levels such as time, temperature, water activity and acidity are meant to monitor the CCP. The limits and tolerances have to be specified for each preventive measure in place and all criteria selected should be documented.

Once again the choice of control criteria and control measures should depend on usefulness, cost and feasibility – but their main purpose is to provide a high assurance of food safety.

Establish a monitoring system for each CCP

Monitoring of critical control points is essential to ensure that the specified criteria are met. Depending on the CCP there are many ways of monitoring foods. The selection of the correct monitoring system is an important part of any HACCP study and may include:

(1) describing the methods by which management can ensure that all CCPs are under control;
(2) detecting any deviation from the established criteria;
(3) detecting loss of control at a CCP in time for corrective action to be taken to regain control; and
(4) providing a continuous recording of criteria such as temperature, pH and time.

Establish a monitoring system

In an ideal situation, a monitoring system should give an on-line continuous picture of performance. In the practical situation, however, the choice of monitoring systems available may often be quite limited. It has to be taken into account that off-line monitoring systems are carried out *away* from the production line which requires extra time before results are available and action can be taken. Whichever monitoring system is chosen – the team must ensure that the results obtained are directly relevant to the particular CCP.

This also means the designation of responsible persons (not necessarily members of the HACCP team):

(1) to carry out the monitoring **(WHO);**

(2) to specify the frequency of monitoring at the CCP **(WHEN);** and
(3) to be in charge of the type of monitoring being carried out **(HOW).**

Everybody involved in the monitoring process must be trained to understand every function thoroughly.

Establish corrective actions

Whenever the monitoring indicates that a deviation from a defined target level has occurred at a particular CCP to a point where the CCP is out of control or in danger of going out of control, corrective actions have to be taken. This means that a specific action plan should be in place whenever a process comes to a point where the established criteria are not met. The specific action will depend on the process being monitored and should, where possible, take place at a point where the product still can be "restored" – in order to prevent the discarding of the product. The actions to be taken and the persons responsible for taking these actions should be clearly established and written down as part of the procedure.

Establish corrective action

Verification

Every HACCP system should be carefully reviewed to make sure that the system is more than just a paper exercise. The HACCP team has to verify that the designed system is working correctly. Therefore, the team has to check regularly whether the following items are still up-to-date and adequate:

(1) flowchart/flow diagram;
(2) significant potential hazards;
(3) CCPs;
(4) specification of the criteria for control;
(5) monitoring procedures to be applied at each control point; and
(6) actions to be taken when the operation is out of control.

In addition, verification should be repeated periodically to check the effectiveness of the monitoring.

This type of verification may include:

(1) checking records of time/temperature readings;
(2) observing operations at critical control points;
(3) taking measurements to confirm the accuracy of the monitoring;
(4) collecting samples;
(5) conducting special studies, for example, inoculated pack or challenge test, with regard to the safety of products; and
(6) interviewing staff about the way they monitor critical control points.

(F Bryan)

It is very important that nothing in the flow diagram, process or composition of the food should be changed without reference to, and possible amendment of, the HACCP plan but if such a change has mistakenly been made then the review process should be able to determine this and allow appropriate remedies to be introduced.

On-site verification of flow diagram

Establish record keeping documentation

This last point is essential to the successful application of a HACCP system. Record keeping also helps the manager to demonstrate at any time that the HACCP principles have been correctly applied. And there is more to it – proper documentation may help in the construction of a "due diligence" defence and is very useful for training purposes.

To make documentation accessible, all collected data should be recorded in a dedicated manual or data system including ingredients, processing data, packaging data, storage and distribution data, deviations and modifications to the HACCP system. Therefore, documentation has a double significance: regular recording helps to check whether the system is successfully working and it also helps management to demonstrate awareness and commitment to the principles of HACCP.

Record keeping helps the manager to demonstrate that the HACCP principles have been correctly applied

5 Hazard analysis for small businesses

So far we have discussed what HACCP is, where it originated and what its advantages are. We have defined its terms, explored its relationship to quality and outlined what the law requires. We have also explained how manufacturers apply classical HACCP using the decision tree.

The aim of this chapter is to show how smaller businesses, including bakers, butchers, grocers, caterers and small processors, can comply with the law, produce a safer product and understand the principles underlying HACCP without going through the process of implementing a classical HACCP system. However, it is not possible to apply HACCP principles to any operation unless the following criteria are fulfilled:

(1) all staff have received appropriate training in food hygiene, including their role in control and monitoring;
(2) the management and staff are committed to "thinking hygienically"; and
(3) the business is operated hygienically.

There is no short-cut to good hygiene

There is no shortcut to good hygiene, if you cannot satisfy these basic criteria then you should take steps to do so before proceeding any further. If you believe you can satisfy the criteria, the next step is to look more closely at the five "principles of HACCP" defined in article 3.2 of The Food Hygiene Directive (93/43/EEC).

The five principles of HACCP

(1) Analyzing the potential food hazards in a food business operation

This means listing all the potential hazards in your business. However, in practice, for the sake of simplicity, hazards in small businesses can be considered in the following groups:

a) microbiological hazards
   i. contamination before delivery;
   ii. contamination after delivery;
   iii. survival of microorganisms; and
   iv. growth of microorganisms.

b) chemical hazards
   i. contamination before delivery;
   ii. contamination after delivery; and
   iii. use of poisonous materials.

c) physical hazards
   i. contamination before delivery; and
   ii. contamination after delivery.

We can already begin to see a pattern emerging, in each case there is the problem of contamination, both before and after delivery.

(2) Identifying the points in operations where food hazards may occur

Let us look a little more closely at contamination, survival and multiplication of microorganisms and the use of poisonous materials.

Contamination before delivery

As mentioned above this can be microbiological, chemical or physical and the risk of contamination will obviously vary from product to product. The risk of physical contamination by, for example, soil, seeds, pieces of wood, stones and insects, is obviously much greater in fruit and vegetables than in, say, meat or fish. Conversely, the risk of microbiological hazards is much greater in raw meat than in fruit. Chemical hazards can vary from pesticide residues in fruit and vegetables to veterinary residues in meat, as well as accidental contamination with chemicals, for example, cleaning materials and chemical reactions in products, typically fish, due to deterioration during bad storage conditions.

Consumer surveys indicate that the greatest public concern about food hazards lies with chemical residues and food additives even when they are used legally. Information from food companies, however, shows that up to 40% of all food complaints from the public are about physical contamination – probably because physical contamination is usually much more obvious than the other types of contamination. This highlights the difference between "actual risk" and "perceived risk". Probably the greatest risk to human health from food is microbiological hazards (see table "Causes of Food Poisoning" page 10) yet the public perception is that the risk from chemicals and foreign bodies is much greater.

It is important, therefore, not only to give equal consideration to all three types of hazard but also to remember that microbiological hazards are the "first among equals".

You must therefore look carefully at any food or ingredients which you buy for your business and note the possible hazards which may arise from them. This includes all raw products, pre-cooked products, water supply and packaging materials. The following table lists some of these and some possible hazards associated with food and ingredients.

Item	Hazards
Raw vegetables	Microbiological. Soil. Small stones. Unwanted plant material. Insects and larvae. Insecticide residues. Fertiliser residues.
Prepared vegetables	Insecticide residues. Fertiliser residues.
Cooked vegetables	Microbiological. Insecticide residues. Fertiliser residues.
Fruit, including fruit concentrates and preserves, and nuts	Microbiological. Moulds. Mycotoxins. Insects and larvae. Twigs. Unwanted plant material. Insecticide residues. Wax coatings. Fertiliser residues.
Cereals	Microbiological. Unwanted plant material. Insects and larvae. Rodent droppings. Insecticide residues. Fertiliser residues. Rodenticides. Moulds. Mycotoxins.
Spices	Microbiological. Soil. Insects and larvae. Rodent droppings. Insecticide residues. Rodenticides.
Herbs	As for raw or prepared vegetables.
Raw meat	Microbiological. Parasites. Faecal contamination. Veterinary residues.
Cooked meats	Microbiological. Veterinary residues.
Raw milk	Microbiological. Heavy metals. Veterinary residues. Cleaning materials.

Item	Hazards
Pasteurized, UHT, sterilized milk	Veterinary residues. Cleaning materials. Heavy metals.
Dairy products made from raw milk	Microbiological. Heavy metals. Veterinary residues. Cleaning materials.
Dairy products made from heat-treated milk	Microbiological. Veterinary residues. Cleaning materials. Heavy metals.
Fish	Microbiological. Toxins. Parasites. Heavy metals.
Farmed fish	Microbiological. Parasites. Veterinary residues
Shellfish and crustaceans	Microbiological. Toxins. Pieces of shell. Heavy metals.
Water	Microbiological. Small crustaceans. Plant material. Soil. Visible metals. Chemical naturally occurring. Treatment chemicals.
Other liquids	Microbiological. Toxins. Chemical.

Control measures for these hazards will be discussed later but, in general terms, you have no control over what happens to the product before you receive it except by:

a) checking supplier's systems;
b) product specification;
c) acceptance of product at delivery; and
d) checking the product before and during use.

The point where these food hazards may occur is therefore at delivery.

Contamination after delivery

This again can be physical, chemical or microbiological and the risk of occurrence will vary from product to product. Unlike contamination before delivery where the control which you can exercise is either reactive or indirect, in this case you, and only you, have control over what happens to a product on your premises. Examples of potential sources of contamination include:

Microbiological contamination

- cross-contamination from equipment or working surfaces
- pests including birds and domestic animals
- direct contamination from raw food
- drainage systems
- water supplies
- air/aerosols
- people
- moulds.

Product must be protected from contamination

Physical contamination

- unwanted raw material, for example, pieces of shell
- personal items such as jewellery, dressings
- defective equipment/machinery
- packaging materials
- hair and fingernails
- rodent droppings
- insects/larvae.

Chemical contamination

- cleaning materials
- pesticides
- rodenticides
- mycotoxins
- air pollution
- water supplies
- chemical reaction between product and utensils/equipment
- maintenance materials, for example, oil.

The points in your operation where any of these potential hazards may occur will obviously depend on the type of food business but it is a relatively easy exercise to look at your own situation and decide the points where hazards are likely to affect your product. If, for example, you are involved in cooking chickens then it is essential to avoid contaminating the cooked chicken with the raw chicken and you should identify the points where this may occur. There may be a number of such points but they could include the following:

- improper storage
- incorrect production flow
- improper cleaning/disinfection of equipment and surfaces
- poor personal hygiene
- poor design of production area.

Most of these are **avoidable** so that when you do identify them you should take positive action to ensure that they are removed. This allows you to concentrate your efforts where they really matter – on the points where you cannot ensure that such positive action is able to eliminate the possibility of a hazard occurring.

To continue with our simple example, if raw and cooked chickens are stored in separate refrigerators then you have avoided that direct route of contamination and all you have to monitor is that the correct foods are stored in the appropriate refrigerator. However, if you have to use the same refrigerator then you must ensure that it is properly stacked with the raw chicken below the cooked chicken, and you must further ensure that this is regularly and properly monitored. It does not take Einstein to work out that, in the long term, having separate refrigerators is the cheaper and safer option in this situation. This consideration should be given to each point until you are left with a list with which you must deal in some other way.

Summary

- List all the points where hazards may occur.
- Consider, in each case, whether or not the operation can be carried out in a different way, or if you need this specific operation at all, so reducing the number of points that need cause concern, bearing in mind that investment in suitable equipment/utensils will pay you back in the long term.
- List the remaining points for further consideration.

Survival of microorganisms

This means the survival of microorganisms following a treatment which was designed to kill them and can be grouped under three main headings

Heating

This includes cooking, baking, reheating, pasteurization, sterilization and ultra-heat treatment. **It is vital that you know and apply the correct time/temperature combination to kill harmful microorganisms if you are relying on this process to produce a safe product.** If you are applying the wrong time/temperature combination or if the equipment is faulty then you may create conditions which allow the survival of harmful microorganisms.

Disinfection, heat and irradiation are used to destroy bacteria

Cleaning/disinfection (of equipment, work surfaces and machinery)

This process is also designed to kill microorganisms and if not carried out properly can result in the survival of harmful microorganisms. **It is essential to apply the correct detergent and disinfectant at the correct concentration at the correct temperature for the correct time using the correct equipment.**

Irradiation

This is another process which is designed to kill pathogens. It is unlikely to be encountered by the average food business except as an irradiated product such as spices and therefore will not be considered in detail. The process itself would always be a CCP.

Heating processes and cleaning/disinfection are usually CCPs and therefore great care should be taken in their consideration.

Multiplication of microorganisms

Microorganisms rely on a number of factors for successful multiplication, including correct temperature, time, food, moisture and acidity. If we wish to control their growth it is necessary to control these factors, although in many situations this will not be possible. For example, in a food business, food will always be present, although harmful microorganisms prefer some foods to others. Furthermore, unless you are a manufacturer, it is difficult for you to control acidity and moisture content. What definitely is within your control is the time/temperature combination of the food on your premises and therefore it is important to remember that any part of your process or business which allows high-risk food to be at any temperature between 5°C and 63°C can allow multiplication. High-risk foods can be defined as those which, under favourable conditions, can support the multiplication of harmful micro-organisms and which are intended for consumption without further treatment which would destroy such organisms. Whilst it is impossible to give a complete list of such foods, they are normally high-protein foods and would, include cooked meat and meat products, milk and milk products, gravies and cooking sauces.

The points to be identified are therefore:

a) where high-risk foods are, for any reason, between 5°C and 63°C;
b) where foods are being kept hot, above 63°C, prior to serving; and
c) where foods are being kept at refrigerated temperatures.

Use of poisonous materials

We have already dealt with chemical contamination where the difficulties are either accidental contamination with chemical substances or publicly perceived risks from chemical preparations or additives. What is meant by the use of poisonous materials is the deliberate but mistaken use of food which contains substances harmful to human health. The types of food normally associated with this type of poisoning are mushrooms, toadstools and fish. One poisonous fish which is eaten quite deliberately is the Fugu fish which is regarded as a great delicacy by the Japanese. This fish, in spite of being prepared by highly-trained chefs, is responsible for many deaths every year in Japan thus leading us to suppose that the flavour is so good that the risk is worth taking! Normally, however, restaurants have no desire to play Russian roulette with their customers and would wish to avoid any foods of doubtful origin or about which the identification is suspect.

Unfortunately, some common foods, unless properly selected, stored, prepared and cooked, can also be responsible for serious chemical food poisoning. Red kidney beans contain a toxin which is only destroyed by boiling. Never eat or serve raw red kidney beans and always ensure that the beans are well boiled before making the chilli con carne! Nutmeg, if used in excess, can give LSD type symptoms caused by a substance called myristicin.

Potatoes are part of the staple diet but the "green" parts contain a substance called solanin which is poisonous; fortunately, solanin is water soluble and therefore the real danger lies in baked potatoes. The sensible message is simply to avoid the green ones. Potato sprouts contain the same substance, therefore it is important to cut off the sprouts.

Shellfish and crustaceans, apart from possible microbiological problems, are normally quite safe if collected from clean waters. Some European and North American waters are, however, seasonally affected by algae which accumulate in the fish and can cause shellfish poisoning. The waters are monitored to check for the presence of these algae, and bans are placed on harvesting, where necessary. Always make sure you know the origin of any shellfish and crustaceans which you use.

Mackerel, tuna and other fish of the Scombroidae species, when raw, must be kept below 4°C, otherwise they can produce a toxin which is associated with high levels of histamine.

Moulds and other primitive fungi that grow on food under poor storage conditions can produce metabolites which are toxic to man and animal. They are called mycotoxins. They can cause cancer and other diseases. The illness can be acute or chronic. Foods usually involved are fruit, cereals and nuts which have become mouldy due to prolonged storage or high humidity. The best approach to prevent contamination from mycotoxins is good storage of susceptible products under clean, ventilated conditions. It is also important to know about the supplier's sources, especially with grains where good farm management is essential.

By now you should have identified the particular points in your operation at which hazards may occur and be in the position to decide which of these points are critical to food safety.

(3) Deciding which of the points identified are critical to food safety – the critical points

We have already decided that, in most cases, cooking and cleaning/disinfection are points which are **"critical"** to food safety but we must look carefully at all the other points which we identified and decide whether or not they are **"critical".**

In order to explore this let us construct some simple flow diagrams for different types of food businesses.

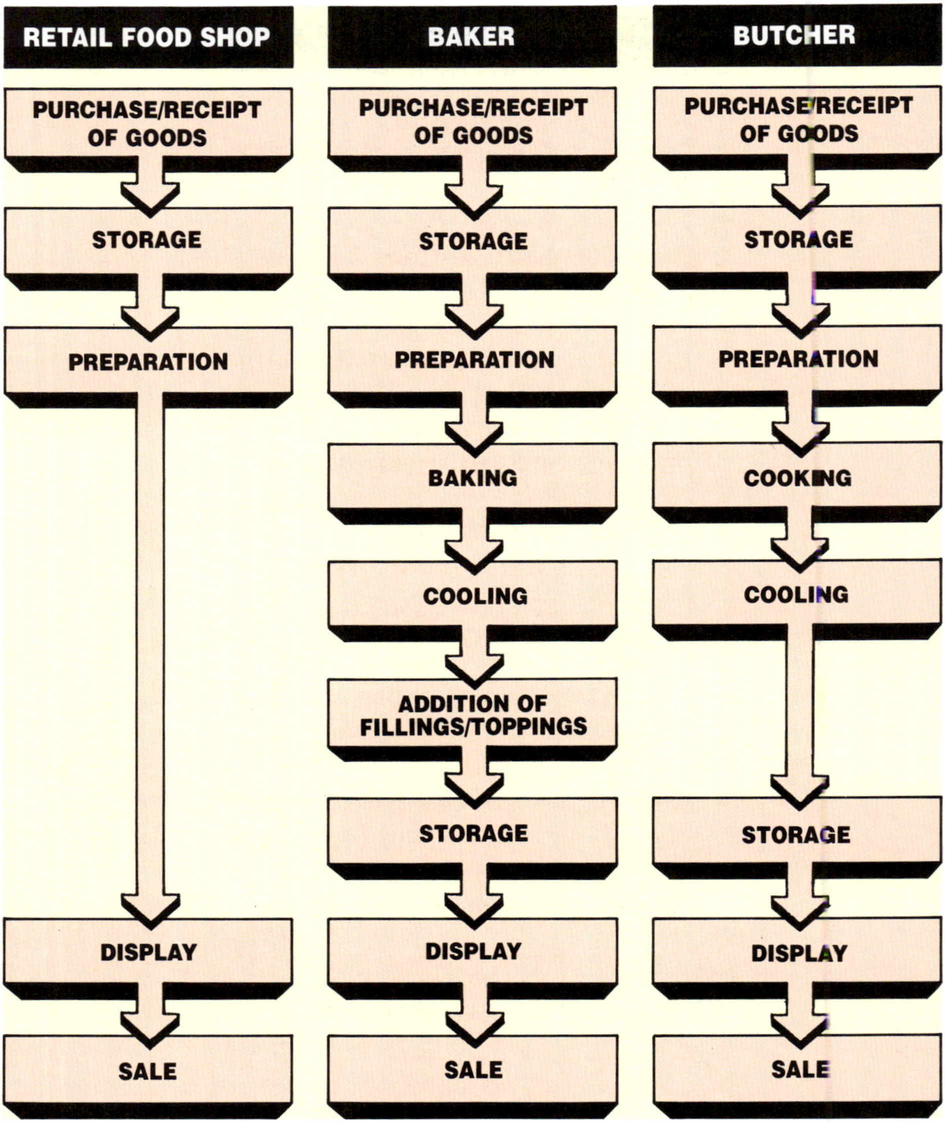

As you can see, many of the steps in each of these flow charts are identical even if the products being dealt with may be different. Although we have included cooking as an activity which may be carried out by the butcher, we have not yet included a flow diagram for **catering.** This is a bit more complicated as the following flow chart shows.

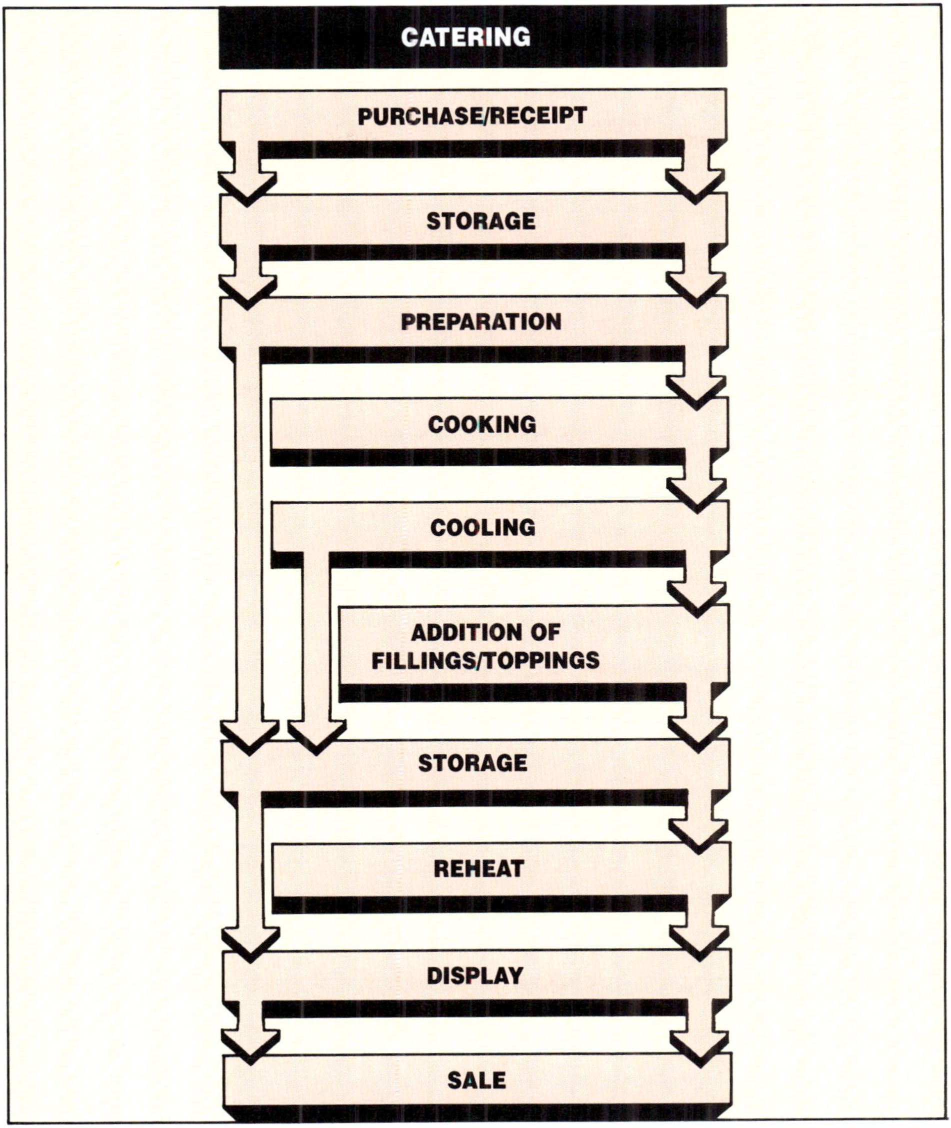

Let us have a closer look at the four flow diagrams above. There are certain steps which are always going to be in a flow diagram no matter what type of business: delivery and receipt of goods, storage, preparation, display and sale. Let us see how it would all fit together in one chart.

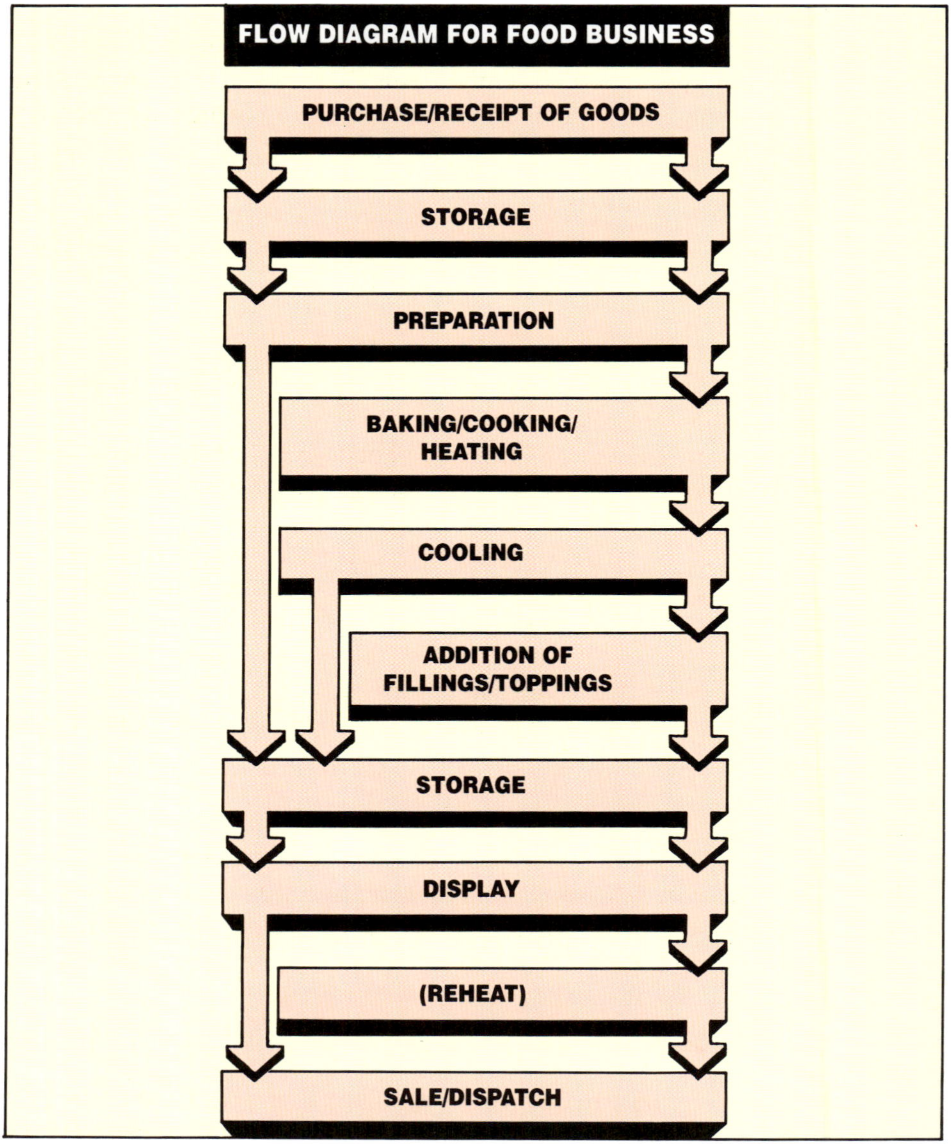

This diagram can therefore be used by most food businesses provided the appropriate route is followed. By using this chart, how can we decide which points are "critical" to food safety?

Purchase and receipt of goods

We have already considered some of the potential hazards found in deliveries of goods – but is this a critical point? In the large manufacturing industry this is a subject of much debate, the details of which we need not discuss, except to say that they are better equipped to test and monitor all incoming materials than other types of food businesses.

However, the types of incoming goods can vary enormously from, say, flour to bread rolls or from milk to ready-made ice-cream and it seems sensible to always consider that this is a critical point in smaller businesses for the following reasons:

a) most small businesses do not have the technical back-up to test the incoming goods themselves;
b) having tests carried out on all incoming goods by an external body would be an expensive exercise and, therefore, any such testing should be kept to a minimum;
c) even if you are further processing these goods you would wish to avoid, wherever possible, being put in the position of having to, unnecessarily, "clean up" contaminated raw materials;
d) it is necessary to have incoming goods which are not going to cause you a problem at some later stage; and
e) you do not want to take up valuable time doing unnecessary checks over and over again.

Storage

If we look back to the paragraphs dealing with "contamination after delivery", it is apparent that many of the factors which lead to contamination are directly related to storage conditions. In addition, there is the problem of multiplication of microorganisms due to storage at incorrect temperatures. Let us look at these two issues separately.

Contamination

In storage, most of the potential for contamination comes from pests, pest control, moulds, cleaning materials and microorganisms. Some of these factors will not be eliminated from the final product even if it is going to be further processed on the premises. For example, mouse droppings in flour will end up as mouse droppings in bread and, similarly, mouse poison in flour will end up as mouse poison in bread and, thus, in the customer. Similarly, chemical taints and tastes from cleaning materials, once there, are there to stay. In the case of bacterial contamination it could be argued that subsequent cooking would destroy any bacteria, but why create a possible hazard when it is just as simple to avoid it?

The argument for making this a critical point becomes even stronger if there is to be no further processing, because any contamination will always remain. Therefore, for the avoidance of contamination, storage should always be considered a critical point.

Multiplication of microorganisms

We have already discussed this in some detail and come to the conclusion that proper temperature control in high-risk foods is essential to prevent the multiplication of microorganisms and it would be difficult to present a reasonable argument which suggested that this was not a critical point.

Preparation

The preparation of food can be divided into two broad categories: preparation prior to baking/cooking/heating and preparation prior to storage/display, and these need to be considered separately. It is important to remember that this is a point in the process where it is almost certain that the food will be "handled" and therefore personal hygiene is of the utmost importance.

Preparation prior to baking/cooking/heating

This can involve a number of activities ranging from the removal of packaging to the mixing of ingredients to the thawing and/or mixing of frozen ingredients, with several possibilities for contamination. Again, it could be argued that any microbiological contamination will be eliminated in the oven, grill pan or pasteurizing plant but why introduce an avoidable risk? In addition, the thawing of frozen food offers the ideal circumstances for cross-contamination and bacterial multiplication.

However, as in storage, it is impossible to remove taints from chemicals and cope with the aftermath of poor pest control. In addition, there is the possible human contribution of such things as hair, fingernails, jewellery and dressings to contend with. Therefore, this has to be a critical point.

Preparation prior to storage/display

The situation here varies from the above only in that it is worse. There is no safeguard of cooking, baking or heat treatment. Another obvious critical point.

Baking/cooking/heating

We discussed earlier the importance of heating processes designed to kill harmful microorganisms and all that needs to be said here is that these processes are always critical.

Cooling

You may recall from the earlier chapter on risk assessment that the commonest factor in food poisoning outbreaks was "inadequate cooling and hot holding" and that this was implicated in 63% of all outbreaks in the UK, USA and Canada. We also discussed earlier the opportunities for multiplication of harmful microorganisms at temperatures between 5°C and 63°C. Furthermore, cooling also gives the opportunity for contamination unless the food is adequately protected. Once again we find ourselves at the mercy of a time/temperature combination which unless properly controlled could cause us serious problems. It does not require anything more than a quick glance at the first sentence of this paragraph to decide that this point is definitely critical.

Addition of fillings and toppings

Many of the most delicious foods have a sauce or a cream filling, or something similar, added after the cooking or baking has been completed. Unfortunately, that means that all the good work which has been done up to this point in identifying and monitoring CCPs could be undone in the final process. Most sauces and cream fillings are ideal growth media for bacteria and therefore have the possibility to contaminate the product as well as support the multiplication of bacteria.

It is important then that the material itself should arrive at this point of the process in as good a condition as the rest of the product to which it is being added. The critical points at delivery, storage, preparation and, if carried out, cooking also apply to the fillings. If contaminated product is introduced at this point or if the process is carried out unhygienically then there is no further point in the process which can eliminate the hazard. It follows therefore that this point is also critical.

Storage and display

In many cases, for example, bakery display units, the display for sale or consumption will be the final storage so it makes some sense to treat these as one, in that the same things can be said about each of them. Furthermore, not all foods are cooled before storage or display – some, in catering or bakeries, are displayed and sold hot (see below). In these cases it is essential that the food is stored and displayed above 63°C.

We have already discussed the importance of cross-contamination, temperature control and physical and chemical contamination and they equally apply here as in the other situations. Again these points are critical.

Reheating

Reheating of previously cooked and chilled foods can be done in a variety of ways, including conventional ovens, convection ovens, microwave ovens and combi-steamers. It is vitally important to thoroughly reheat the food until the temperature at the centre (the core temperature) has reached at least 75°C or such higher temperature as the law might require (82°C in Scotland). Reheating should be done as quickly as possible in order to prevent multiplication of bacteria. This is yet another "critical" point because there is no further step in the process which will kill any bacteria which are present.

Sale/despatch

If you are despatching goods which require to be temperature controlled then this is simply "storage on the move" and should be treated no differently to any other kind of storage; critical until it reaches your customer. Additionally, any potential physical or chemical hazards would have to be considered.

At the point of use, whilst you remain legally liable for the safety of the food which you have sold, what happens to the food is out of your control and provided you have dealt with it safely and given adequate instructions to your customer about storage, cooking and heating then, at last, you have reached a point which is not critical.

(4) Identifying and implementing effective control and monitoring procedures at those critical points

Control

We discussed in the previous section how to decide which points in an operation were critical, i.e. **critical control points,** but having made these decisions what do we do about it? All we need to do is establish which hazards we are trying to control, how we are going to control them and what tolerances we are going to allow. We have already looked at the hazards, all we need to do now is look at some control measures and tolerances.

The table in Appendix III sets out a list of control measures but let us look at one or two in detail. It is best to start with an easy one, cooking. For the moment we can set aside the fact that most cooking is done to make food palatable, and concentrate on cooking being a process to make food safe. The purpose of cooking is, therefore, to destroy pathogenic bacteria, i.e. cooking is a control measure. What tolerances should be set? It could be argued that since no tolerances should be set that would allow the survival of pathogenic bacteria, then if the time/temperature combination is considerably in excess of this there is no need to control or establish tolerances or monitor time/temperature – **wrong.** If, for example, the combination required for the safety of a particular product is an oven temperature of 90°C for 30 minutes and the actual time/temperature at which the product is cooked is 120°C for 45 minutes, then you would have no way of knowing that either the time or the temperature had dropped below the critical level unless tolerances had been set around 120°C/45 minutes. An appropriate tolerance would therefore be in the region of 115°C/40 minutes. You could then be sure that not only was the product safe but also that it was cooked in the way you wished! This is a fairly straightforward example because the difference between the safe level and the actual level is so great that setting the tolerance limits does not require a great deal of skill. It becomes much more difficult to set the limits, and as we will see later to monitor them, the closer together these levels are. The simple rule is to have as large a margin of safety as possible.

Another control measure is the receipt of goods. There could be many criteria to set for this but, to use a simple example, for chilled or frozen foods the delivery of the goods at the correct temperature is always going to be critical. The actual temperature of chilled goods could be a legal requirement, in which case the tolerances are already laid down or if it is not a legal requirement then it should be detailed on your product specification. For example, if your product specification requires the temperature at delivery to be below 5°C then you could decide to have no tolerance, a one degree allowance or two degrees and so on but it is unlikely that you would wish to accept more than a couple of degrees because it would then start to negate the benefits of the original temperature. In other words the tolerance in this case is the upper limit which you are willing to accept.

Another control measure frequently quoted for raw materials is “only buy from a reputable supplier.” Whilst this, at first sight, seems a sensible piece of advice, what exactly does it mean? How do you know whether a supplier is reputable or not? Some might answer that the supplier would be reputable if operating a Quality Assurance scheme such as ISO 9000 but the quality in such systems is only to the standard that is built into the system. In other words, the QA system merely ensures that all goods are turned out consistently to the defined quality but that quality may be lower or different to that which you require. The best way to ensure that you receive the goods you want is to give your supplier a product specification which lays down your requirements, and to check that your specification is being met. If you then find that your product specification is adequately covered by the supplier's QA system then you could be reasonably confident that this is a reputable supplier.

Once you have decided what you wish to control and what tolerances you wish to set then you must decide how to monitor them.

Monitoring

Broadly speaking there are two types of monitoring:

a) those criteria which can be monitored on the spot, for example, times, temperatures, physical appearance, pH, foreign bodies, sensory values, training and competency of staff; and
b) Those criteria which require laboratory analysis, for example, microbiological counts, chemical composition and residues.

It is obviously cheaper and quicker to have as many as possible of the control points under category (a) and as few as possible under category (b) although laboratory analysis is sometimes unavoidable. The other obvious disadvantage of laboratory monitoring is the time taken to obtain the results, which can often be several days. This may not be such a problem when monitoring incoming raw materials because it may be several days before they are used, but is certainly an insurmountable problem if it is required several times during the process run.

What kind of monitoring can we do on the spot? Let us look briefly at some monitoring which can be done at different points of our flow diagram.

Purchase/receipt

As discussed above, the temperature of chilled and frozen goods can be checked. A whole range of physical checks can be carried out, including smell, taste, spoilage, freshness/date codings, physical damage, moulds and so on, depending on the type of product.

Storage

Arguably the most common hazard in storage is pests: insects, rodents and birds. It is quite simple for you to check for signs of infestation on a regular basis and, as at all stages, good pest control is essential. Other hazards are contamination by chemicals and in this case a simple, but regular, visual check will suffice to ensure that all chemicals are stored out of harm's way. This also applies to problems of cross-contamination where, for example, a visual check will ensure that cooked and raw foods are being stored separately or at least in a way that prevents a problem.

Preparation

Again the ever present spectre of cross-contamination must be dealt with and proper separation of products and a good cleaning schedule are essential. Separation can be built into the system and should not be a problem if everyone is aware through training and supervision of the importance of this factor in producing a safe product. Likewise, a good cleaning schedule is easy to follow and monitor if there is a will to do so. This also applies to such things as jewellery, hair and other human contributions. More difficult, perhaps, is the monitoring of the time taken to completely thaw frozen products, particularly poultry and joints of meat, as this can vary according to weight, thickness, type of product and even the time of year and it must always take place in hygienic conditions; not an impossible task by any means but one which requires careful thought to ensure that the thawing process in itself does not introduce a hazard by, for example, cross-contamination with thawed liquid. It is also important to ensure that complete thawing takes place, otherwise during the cooking process the heat is utilized in thawing rather than cooking leading to undercooking and risk of food poisoning. The following table taken from Richard Sprenger's "Hygiene for Management" illustrates some approximate thawing times for frozen poultry.

Thawing times of frozen poultry

Oven ready wt. Kg (lbs)	Approx. thawing time at room temp. (hours)
2.25(5)	15
4.50(10)	18
6.75(15)	24
9.00(20)	30

Cooking/baking/heating

As discussed before, it is the time/temperature combination which is important and requires monitoring. The use of a properly calibrated oven thermometer will ensure the correct oven temperature and the core temperature (temperature at the centre of the product) can similarly be measured by a calibrated probe thermometer. It is important to measure the core temperature as it always takes longer to heat the centre of a product than the outside. The measurement of time requires nothing more than a reasonably accurate clock!

Other heating processes such as pasteurization employ more sophisticated techniques as the time and temperature need to be accurately measured and use extra safeguards such as flow-diversion valves and thermographs.

Cooling

If cooling is to be a CCP then monitoring it is not really an easy business. Naturally, it should be carried out in hygienic conditions and as quickly as possible, but measuring the temperature is not a simple matter because it is fairly pointless to take a single measurement as this only gives information at that point in time and is thus of limited value. Of equally limited value is the temperature in the thawing cabinet as this does not indicate the temperature of the food or its cooling rate. It is much more relevant to measure the temperature before cooling and at regular intervals until the whole product has reached the required temperature. It also helps to have joints of meat as small and thin as possible and trays of food as shallow as possible in order to speed up the cooling process. Additional monitoring could include checking that the food is covered and that there is sufficient air space above, below and around the product.

Addition of fillings/toppings

At this point, monitoring could include ensuring that both the food and the filling or topping are at the correct temperature, that the operation is carried out hygienically and that filling/topping is only removed from storage in the quantities which are immediately required. Good personal hygiene, always important, is essential at this point because many of the operations are carried out manually.

Storage/display

Storage and display can involve hot holding, chill storage and ambient storage. The hot holding temperature should always be above 63°C and any monitoring should be designed to ensure this; we have already discussed methods of monitoring temperatures. Cold holding temperatures depend on the product and any legal requirements, but as a general rule should not be above 8°C. Of equal importance is the date marking of foods which allows a proper system of stock rotation to be observed.

High standards of personal hygiene are essential to avoid food contamination

Reheating

It is vitally important to get the time/temperature combinations correct otherwise this step in the operation could result in serious consequences. You must be able to have confidence that the food has been reheated to the desired temperature as quickly as possible. This can be done in a number of ways, from the simple use of a probe thermometer through to the technical reliability of a combi-steamer.

Sale/despatch

All that needs to be done at this point is to ensure that the cool chain is not broken, that the goods are not contaminated by staff or customers and that your customer benefits from the care which has been given throughout your process to ensuring that they receive a safe, quality product.

Corrective action

Essential to the control and monitoring procedures is laying down the corrective action to be taken if the monitoring shows that the system is operating outside the tolerances. In other words, it is pointless to set up a wonderful system to ensure safety if nobody knows what to do if the cooking temperature is not reached or the refrigerator is running at room temperature or the haddock smells of petrol.

It is essential that the staff in charge of monitoring are properly trained to know what to do or whom to contact if the system is operating outside the tolerances.

Target levels and tolerances must be established

(5) Reviewing the analysis of food hazards, the critical control points and the control and monitoring procedures periodically and whenever the food business operations change

There are a number of reasons for reviewing the system. It is necessary to check that not only are the CCPs being properly monitored but that they are proving with experience to be the correct CCPs. **It is essential to review the system when any part of the procedure is changed, for example, equipment, ingredients, temperatures, production flow, suppliers, shelf-life and cooking times, because the original decisions may have been made invalid by the changes.**

Changes should only be made to the system when due consideration has been given and everyone involved is acquainted with them.

6 Last but not least - A word on documentation

We have already mastered the jungle of definitions. We have climbed the staircase of the numerous principles of HACCP and have noticed with some relief that the EC Directive offers a short-cut by quoting just five necessary steps to take. Studying these five principles we find out that one of the key words of HACCP is not quoted or even recommended: **documentation.**

But, having made the effort to apply HACCP principles to your business, it would be a lost opportunity if some of the procedures were not written down. In fact some people would argue that the successful application of HACCP principles depends on some form of documentation, no matter how brief or basic. Furthermore, this written documentation is ideal for communicating the requisite hygiene standards to staff, is extremely useful for training purposes and will prove invaluable when the HACCP system is being reviewed or when a specific control measure does not appear to be working satisfactorily.

HACCP is not about licking your finger to make out the direction of the wind. Applying HACCP principles means complying with the law, and by demonstrating that you have done everything to ensure food safety it goes a long way to providing a due diligence defence. This can be proved by the correct documentation, and it is your documentation that will count with the enforcement officer who visits your premises.

Records are written evidence which document a process that has taken place and should be available for inspection whenever required. **Remember, however, that the law does not require you to keep any records.** But should you decide to adopt a sensible approach, let us consider where documentation is necessary and what it should include.

Documentation of HACCP principles

"Analyzing the potential food hazards in a food business operation" means finding out the hazards specific for product as well as process. Therefore, documentation at this step has to cover both the product and its processing.

PRODUCT
DESCRIPTION AND DEFINITION
product specification list of raw ingredients list of pre-processed ingredients and their method of preservation detailed composition

PROCESS	
MAIN	**SUBSIDIARY**
flow diagram of operation (general) complete processing record including storage and distribution	cleaning and disinfection: schedules and substances used pest control schedules and substances used packaging (if not incorporated in line)

The critical points and their documentation

We have identified the critical points in food operations. The critical points should then be listed, preferably in a flow diagram document. In this context every operational step should also be described according to the hazard present at the point together with its control measures. Let us have a look at a classical example of a critical point: the process of heating milk.

Operation – heating of milk	
hazards	pathogen spores survive heating; vegetative pathogens survive inadequate heating
critical point?	YES
control actions	pasteurize or boil/sterilize
monitoring procedures	time/temperature control
target level and tolerances	according to the chosen method

As the specific target levels and tolerances for each critical point represent some measurable parameter (time, temperature, moisture, pH, a_w, etc.) they can easily be documented. We have already seen that monitoring systems can be continuous (for example, temperature control on thermographs), or at a point in time (sampling).

A corrective action plan is necessary if a critical point has deviated from its specified tolerance

Last but not least, every critical point requires a corrective action plan. When monitoring results show that a critical point has deviated from its specified tolerance, a corrective action plan has to be at hand. It is also necessary to have a plan for the disposal of food that has been produced during the critical time. Let us have a look at another example, the frozen storage of meat.

Critical point – frozen storage of meat	
hazard	microbial growth in thawed food
control measures monitoring procedures	maintain frozen until use. Observe whether goods are frozen, continuous temperature control of freezer, alarm system (acoustical/optical) in case of temperature increase
target level	temperature of frozen meat –18°C
tolerances	temperature in marginal layers –15°C (short term)
deviation	temperature > –15°C
corrective action	immediate thawing and further processing of meat or disposal

Note that whenever a corrective action has to be taken, the established flow diagram has to be changed to ensure the safety of the end product. It is obvious that this has to be documented. Both general monitoring and corrective action (and disposal action) has to be incorporated in the HACCP record keeping.

The documentation of monitoring

This is probably the most important part of documentation because this is the daily recording of your safety concept. Some monitoring methods are recorded automatically but, especially in small business, many parameters have to be checked personally (for example, correct temperature of refrigerators and correct storage temperature even if it consists just of reading a thermometer scale. It is not only important to protocol the controlled parameters (and their deviations) but you should go one step further – you should check your monitoring, which means writing down:

- frequency of monitoring (daily, weekly, etc.);
- timing (beginning of shift, during the day, end of shift, nightly); and
- person responsible.

If you are running a one-person business, it is all in your hands. If you employ staff, it is your responsibility to check whether the critical points have been monitored by a designated person. This means that you need to look at a report sheet (this can also be a mechanical temperature graph or a computer print-out, depending on the kind of monitoring equipment you use). Sign it and keep it in a special folder to demonstrate at any time that you have been observing due diligence.

Let us now summarize what should be put down:

- What is my product (specification, ingredients, formulation)?
- What is my production (flow chart)?
- Where are the critical points and what are the hazards that need control?
- What especially needs control (parameter, target level, tolerances)?
- How do I look after it (monitoring procedures)?
- What do I do if something goes wrong (corrective action)?

This is your HACCP plan which has to be completed with daily records on monitoring (see above). Let us add just two more points:

- When did I last check "my HACCP" system as a whole, and what did I find out?
- Is there anything that needs changing/improving?
- What have I changed? How did I modify "my HACCP" according to the change?

These last two points are simply "verification" and "reviewing", and their frequency depends on the type of business you are running and the changes you are undergoing in product, processing, premises lay-out, cleaning and disinfection of equipment, etc.

Keep in mind that our list is not necessarily exhaustive. You can always add to your documentation whatever you feel suitable. But the points mentioned above are the minimum which should be documented.

Record keeping, documentation, the whole HACCP plan – it looks like a lot of paper work. This is not necessarily true if it is done efficiently and economically. Documentation, although never mentioned in the EC Directive, is essential to the successful application of HACCP principles. It also facilitates food control because an accurate HACCP documentation will show everybody that you have played your part in food safety.

We have now looked at what the law requires, discussed how the principles of HACCP can be applied to any business, looked at a flow diagram which can be adapted for most businesses and in the appendices you will find examples of flow diagrams and control and monitoring in a selection of businesses.

Now it is up to you. HACCP has, hopefully, become less of a mystery – applying it in your business will not only help you to stay within the law but also save you time, effort, money and worry in the pursuit of a safe product for your customer. To sum up:

the "principles of HACCP" are simply "principles of common sense".

Appendix I HACCP and quality

People often confuse HACCP with Quality Assurance. **HACCP is not a Quality Assurance system** – it should not be confused with ISO 9000 or similar quality systems. Quality foods are not necessarily safe, for example, undercooked foods may be of the highest quality but are unsafe. Conversely, burnt food may be perfectly safe but who would want to eat it? Ideally of course, food should be of good quality and be safe, therefore many businesses which have gone to the trouble of introducing a Quality Assurance system will have incorporated a HACCP system. The point is that you do not need a QA system in order to have a HACCP system. HACCP is about **safety** not quality, although both systems are concerned with consistency.

Although we have already agreed that consumers have the right to expect their food to be safe, most consumers will probably be more aware of the quality of their food rather than its safety. It is important therefore to discuss the concept of quality.

Public opinion research shows that ideas and perceptions of quality differ significantly depending on the expectation of the customer. Therefore – **what is quality?**

The term quality usually implies such terms as "goodness", "worth" or "value"; for example, saying that somebody is aiming at quality rather than quantity implies that they aim to produce superior goods rather than large quantities. Inquiries about food quality can also give rise to a rather large spectrum of responses such as "freshness", "healthiness" and "wholesomeness". It is no surprise, therefore, to discover that the quality of food can be determined by many factors including:

- *sensory value:* look, smell, taste, consistency, freshness;
- *economic value:* potential uses, waste disposal, processing/cooking costs, packaging, storage;
- *health value:* nutrition, minerals, vitamins, hygiene, pollution; and

- *psychological value:* environmental issues, ideals, image/prestige/status and attitudes.

The quality of a product can therefore be defined as:

“The sum of the characteristics of a product having regard to the standards set by the customers’ expectations.”

A quality product is not necessarily the most expensive one – what matters most is its suitability for the required purpose, and therefore a cheap dish that requires little preparation and satisfies the strongest hunger may often be better than a luxury meal.

“To catch a glimpse of quality you have to look at it as a reward and not as a target.”

(A. de St. Exupéry)

Well, this poet’s point of view will not necessarily appease a dissatisfied customer who has been reassured that every staff member has tried really hard not to look at quality as a goal! Therefore let us look at it this way:

“Quality means the return of the customer – not the food.”

The EC Directive on the hygiene of foodstuffs has suggested the use of the EN 29000 series when dealing with good hygiene practices. According to this series, quality means conformity with specific characteristics or features resulting from needs brought to light by public market inquiries. Furthermore, quality within the EC regulations is subjected to guidelines and recommendations for producers. What then is Quality Assurance? and what has quality to do with HACCP?

Again we have to make clear that the implementation of HACCP means **the assurance of product safety.** On the other hand there is also a need to ensure a constant level of quality. This is called **Quality Assurance.** Quality Assurance is a concept which seeks to ensure that a product or a food service will continuously satisfy the customer’s quality expectations.

The concept of “thinking things through” applies to Quality Assurance as well as to HACCP. Like HACCP, Quality Assurance is proactive as distinct from **Quality Control** which usually takes place at the end of the process chain via end-product testing. This means that Quality Assurance attempts to stop things going wrong in the first place. To be fully effective, Quality

Assurance should, as discussed earlier, incorporate the principles of HACCP, as **both systems** can be seen as tools in hygiene management and both are concerned with consistency.

But the practical implementation of Quality Assurance is another complex subject beyond the scope of this book. What we should keep in mind is that quality does not imply product safety, but that the principles of HACCP can be used to ensure product quality. Let us finish this "intermezzo" on quality with one more quotation, this time from Richard Sprenger's "Hygiene for Management":

"Quality must be built into a food product; it cannot be inspected in."

Appendix II The decision tree

It may be an interesting coincidence that in many religions a tree plays an important part in the attainment of spiritual enlightenment. We have already decided that HACCP is no religion. Philosophically, however, we find ourselves confronted with another tree of knowledge – the decision tree – which turns out to be a tool for identifying the critical points of a process – the CCPs.

"The determination of the critical control point is the heart of HACCP. The keywords for success in this task are flexibility and common sense."
(Bob Mitchell)

The decision tree offers a way to identify points of control **and** to distinguish between control points and **critical** control points. We have already defined a CCP as a point (process, practice, etc.) at which control must be exercised to minimize or prevent a hazard. The sequence of questions building up the decision tree add another aspect.

Considering a hazard at a particular step of the flow diagram, we have to ask the question:

**Q1 Are control measures in place for the hazard? or
Could preventive measures [PMs] exist?**

If the answer is **No,** which means that preventive measures cannot be introduced at this step, then the step cannot be a CCP because it is not possible to effect control. As a consequence, we must ask then if control is **necessary** at this step to ensure product safety. If not, we can conclude that the step is not a CCP. If, however, the answer to this supplementary question is **Yes,** it becomes obvious that this step has to be **modified** so that control can be obtained over the specified hazard.

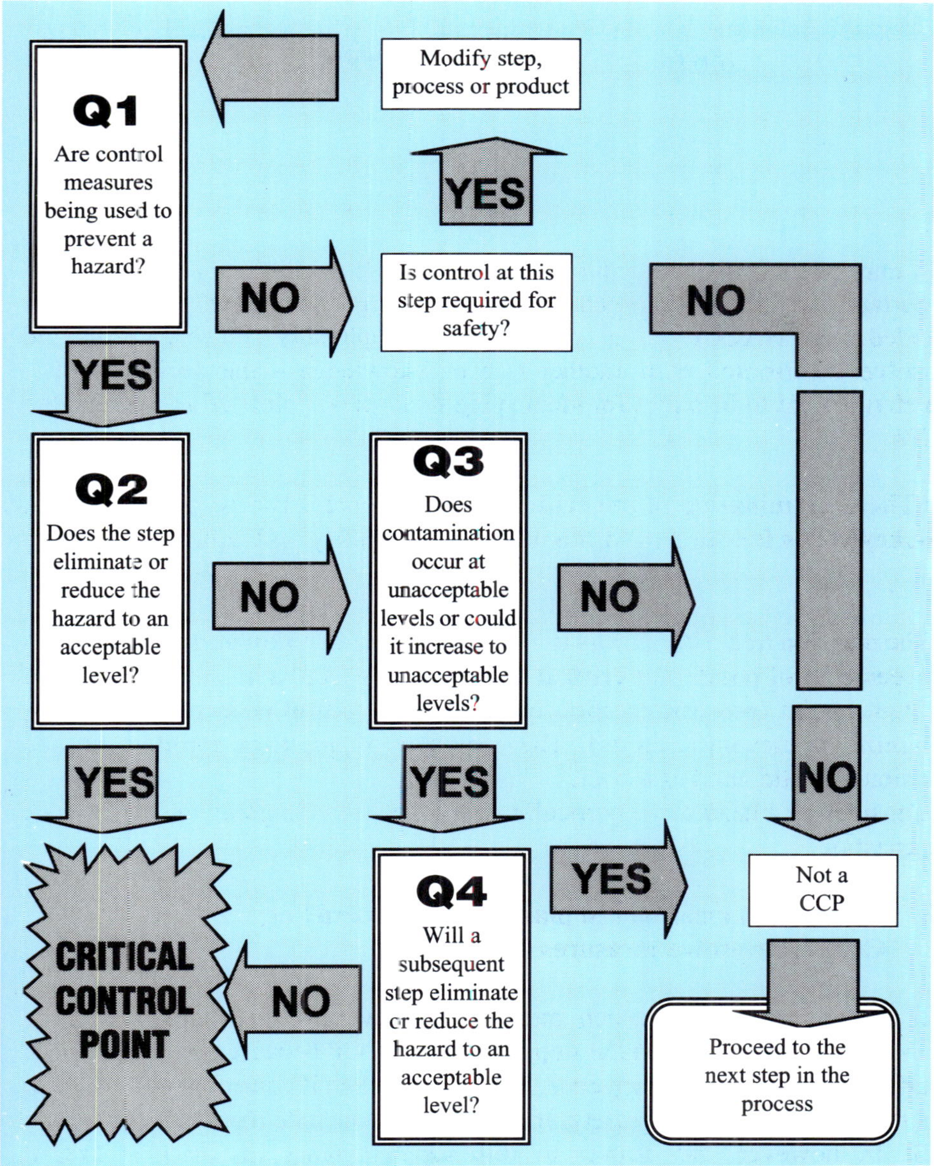
Modify step, process or product
Q1
Are control measures being used to prevent a hazard?
YES
NO
Is control at this step required for safety?
NO
YES
Q2
Does the step eliminate or reduce the hazard to an acceptable level?
NO
Q3
Does contamination occur at unacceptable levels or could it increase to unacceptable levels?
NO
YES
YES
NO
CRITICAL CONTROL POINT
NO
Q4
Will a subsequent step eliminate or reduce the hazard to an acceptable level?
YES
Not a CCP
Proceed to the next step in the process

If the answer to Q1 is **Yes,** we follow the trunk of the decision tree further to consider the next question.

Q2 Does the step eliminate or reduce the hazard to an acceptable level?

This question sounds familiar, and it is indeed nothing else than the description of a CCP. Therefore, if the answer is **Yes,** we know that the process step under consideration is a critical control point. At this point it's essential to identify precisely what is critical (this means taking into account all available data for the hazard under consideration).

If the answer is **No,** we have to carry on to the next question.

Q3 Could contamination occur at unacceptable level(s) or increase to unacceptable level(s)?

In other words, at this point in the process could the product become contaminated with any of the identified hazards to such an extent that it will not be reduced to a safe level or eliminated at a later stage in the process **OR** could microorganisms already existing in the product multiply to such an extent that they cannot be reduced to a safe level or eliminated at a later stage?

It is obvious that if the answer is **No** we do not have a CCP at this step. But at this point, careful consideration of all factors such as ingredients, processing environment and cross-contamination as a likely source of hazard is necessary. Another important aspect not to be underestimated is the possible *increase* in levels of the hazard via a number of process steps. Any team discussing Q3 about a specific process step has therefore to consider not only the step in itself but also the possible accumulated effect of subsequent process steps. In case we are unsure of any aspect of acceptance the answer to Q3 has to be **Yes.** This means we are confronted with one last question.

Q4 Will a subsequent step eliminate or reduce the hazard to an acceptable level?

If the answer is **No** then the step is a CCP. If the answer is **Yes** then the step is not a CCP, as the considered hazard can be brought under control by supplementary process steps.

It is interesting to look at Q4 not as an isolated **"Yes** or **No** question" but in close connection with Q3 and Q2. Proceeding with the decision tree until Q4 means nothing else than the tolerance of the presence of a hazard at a certain process step **if that hazard will subsequently be eliminated or reduced to an acceptable level,** for example, the presence of pathogens in raw material which will be cooked in a subsequent process step. Therefore Q3 and Q4 are designed to work in tandem, otherwise nearly every process step in an operation might be regarded as critical, and we end up with far too many CCPs to be effectively controlled.

Therefore this question also works with Q2. Knowing about subsequent process steps we can accept a contamination at an early stage of the flow diagram with the certainty that it will be eliminated or reduced to a safer level in a following step. For example, Salmonella may be present in raw meat, but if it is destroyed by a subsequent process step, for example, cooking, its presence at this point is not necessarily critical. This reveals another advantage of the sequence of questions: attention is focused on the realistic end use of the product, which stresses the importance of further instructions such as appropriate product labelling for the customer's safety.

How do you work with the decision tree?

The decision tree has to be applied to each step of the process. *All* hazards which may be reasonably expected to occur, or be introduced at each step, have to be considered – this of course means "starting" the decision tree at each point over and over again.

The identification of CCPs

Appendix III

Critical control points

Step	Hazard	Control	Monitoring
Purchase of raw materials	Presence of contaminants; mirobiological, chemical or physical.	Select least hazardous ingredients. Only use reputable suppliers (see text). Specification for product quality and safety including delivery temperatures.	Inspect supplier or request records to show that they follow good manufacturing practice. Historical check of deliveries. Absence of customer complaints. Bacteriological sampling.
Receipt of raw materials	Presence of contaminants. Multiplication of microorganisms.	Specify delivery requirements, especially time and temperature. Minimize time for unloading/ placing in storage. Deboxing area. Staff training.	Check delivery vehicles and drivers, date codes, time for unloading and temperature and condition of food (as per specification). Staff records and competency testing.
Storage (chilled, frozen and dry)	Multiplication of microorganisms. Contamination due to poor hygiene practices.	Store at correct temperature (alarmed units). Cover/wrap food. Stock rotation/date codes. Separate raw/high-risk foods. Cleaning/disinfection. Proper storage of cleaning materials. Good housekeeping. Safe use of pest control/pest-proof containers. Staff training.	Check air/food temperatures, date codes, pest monitoring and food complaint records. Audits and visual checks of food. Cleaning schedules. Staff records and competency testing. Equipment maintenance.
Preparation	Multiplication of microorganisms. Contamination due to poor hygiene practices and poor personal hygiene.	Prepare minimum amount of food. Minimize time at room temperature. Good personal hygiene/training. Separate raw/high-risk foods and use of separate equipment. Colour coding. Cleaning/disinfection. Good hygiene practices, organization/workflow.	Check time/temperature. Audits/visual checks. Competency testing of staff. Cleaning schedules. Bacteriological swabbing of hands and surfaces. Equipment maintenance. Design. Pest monitoring. Electronic fly killers.
Cooking/ baking/ heating	Survival of microorganisms.	Centre temperature at least 75°C. Observe time/temperature combinations. Ensure frozen poultry/joints completely thawed. Staff training.	Check time/temperature. Equipment maintenance. Audits. Staff records and competency testing.
Cooling	Multiplication of surviving microorganisms or germination of spores. Toxin production. Contamination.	Weight/thickness of joints. Cool rapidly to < 8°C (blast chiller). Cool in shallow trays. Keep covered. No contact with raw food. Pest control. Good personal hygiene/training. Cleaning/disinfection.	Check time/temperature. Audits/visual checks. Equipment maintenance. Staff records and competency testing. Pest monitoring. Hand swabbing. Cleaning schedules.

Step	Hazard	Control	Monitoring
Addition of fillings/ toppings	Contamination, Multiplication of microorganisms.	Use clean and disinfected equipment and working surfaces. Use separate equipment to avoid cross-contamination. Shorten time in critical zone (5°C to 63°C) Good personal hygiene and training. Pest control.	Audits/visual checks. Equipment maintenance. Staff records and competency testing. Pest monitoring. Hand swabbing. Cleaning schedules
Storage	Multiplication of microorganisms. Contamination (microbiological, chemical and physical) due to poor hygiene practices.	Store at correct temperature (alarmed units). Cover/wrap food. Stock rotation/date codes. Separate raw/high-risk foods. Cleaning/disinfection. Proper storage of cleaning materials. Good housekeeping. Safe use of pest control/pest-proof containers. Good personal hygiene and training.	Check air/food temperatures, date codes and food complaint records. Pest monitoring. Audits and visual checks of food. Cleaning schedules. Staff records and competency testing. Equipment maintenance.
Display	Multiplication of microorganisms. Contamination (microbiological, chemical and physical) due to poor hygiene practices.	Store at correct temperature. Cover/wrap food. Stock rotation/ date codes Separate raw/high-risk foods. Cleaning/disinfection. Proper storage of cleaning materials. Good housekeeping. Safe use of pest control. Good personal hygiene and training.	Check air/food temperatures, date codes. Pest monitoring. Audits and visual checks of food. Cleaning schedules. Staff records and competency testing. Equipment maintenance.
Reheating	Survival of microorganisms.	Reheat to above 75°c. Staff training.	Check core temperatures. Staff records and competency testing.
Sale/ despatch	Contamination (microbiological, chemical and physical) due to poor hygiene practices.	Keep <5°C or >63°C. Keep food covered. Stock rotation/date codes. Separate raw/high-risk foods. Cleaning/disinfection. Proper storage of cleaning materials. Good housekeeping. Safe use of pest control. Good personal hygiene and training. Cleaning/disinfection. Sell within shelf-life. Prevent customer contamination.	Check time/temperature and date codes. Audits/visual checks. Staff records and competency testing. Cleaning schedules. Pest monitoring. Equipment maintenance.

Appendix IV Examples of hazard analysis in small businesses

Small retail grocer's shop

A small retail grocer sells many different items but they all fall into categories or groups, which is very convenient when developing a hazard analysis system. For example, most goods come prepacked in cans, jars, bottles, boxes, bags and have a relatively long shelf-life. These are easily dealt with by simply checking the condition on purchase/delivery, by practising good stock rotation, by storing properly and not selling past the best-before date. Some other goods are bought in bulk, for example, cereals, and are weighed out and packaged on the premises – with these goods, personal hygiene and pest control obviously become more important than with most of the packaged goods discussed above.

A third, and important, group is **chilled** prepacked foods. These are generally high-risk foods with a fairly short shelf-life.

In this case, it is a very straightforward flow diagram, as in the vast majority of shops there will only be two steps: purchase/receipt and display/sale. All that then remains is to have a look at which hazards we need to be concerned about and how to control and monitor them, which can be done by looking at Appendix III.

Small baker's shop

A small baker not only sells but also manufactures a large variety of product lines and at first sight applying HACCP principles to such a business would seem a daunting task. However, in common with grocers, many of the products can be grouped together for the purpose of hazard analysis.

For example, there may be various types of bread rolls, all of which look and taste differently but from a hazard analysis point of view they are essentially the same product with the same flow diagram, hazards, CCPs, control measures and monitoring procedures. It does not really matter if you produce one type of bread or 21 types; the hazard analysis is the same.

This logic also applies to other groups of bakery products such as fruit pies, meat pies, cakes and so on. Arguably, the point of highest risk is the addition of fillings or toppings to cakes and pies. A flow diagram for a cream cake, for example, would be: purchase/receipt, storage, preparation, baking, cooling, addition of filling and display/sale. If we look at Appendix III we can pick out the appropriate hazards and their control measures and monitoring.

Small butcher's shop

Much of the activity in these premises involves the purchase, receipt, storage, display and sale of raw meat in various forms and as such a hazard analysis is fairly easy to do using Appendix III.

Many butchers manufacture meat products but again they can be divided into categories, those that are raw and have to be cooked by the consumer, those that are cooked but require reheating and those that are ready to eat.

In a similar manner to that demonstrated above, flow diagrams can be constructed for each of these operations and the appropriate hazards considered together with their controls and monitoring. Particular attention should obviously be paid to personal hygiene and cross-contamination between raw and cooked products.

Catering premises

Catering is probably the food business with the most diverse and complicated range of products but even here it is relatively easy to apply the principles of HACCP to each step in the operation. The Department of Health in England has published an excellent booklet, "Assured Safe Catering" ISBN 0-11-321688-2, which describes in detail a fairly straightforward hazard analysis system for caterers.

Once again the trick is to group together similar types of products and treat them as if they were the same. In other words, you look at operations such as storage, cooking, hot holding, etc., just as we have done in previous examples and do not try to deal with each product as an individual item. Depending on the type of operation you can however make life less complicated by a change in the process as the following example shows.

Imagine that you are preparing a chicken salad. Let us forget for the moment the "salad" ingredients and concentrate on the "chicken". There are a number of ways you can buy this chicken, fresh or frozen, cooked or raw, whole or portioned. The reasons for your choice may be economic, quality, convenience or practicality, but that choice can also influence your hazard analysis.

(1) *Frozen whole raw chicken.* A flow diagram for this operation would include the following steps: purchase/receipt, storage, preparation (including thawing), cooking, cooling, storage and finally portioning, plating with the salad and serving (selling). There is nothing wrong with this method but it does involve quite a number of steps with their accompanying hazards, controls and monitoring.

(2) *Fresh raw chicken portions.* This is not a great deal different to (1) above but we have eliminated at least two hazardous operations, thawing and portioning. Our hazard analysis has therefore changed quite dramatically merely by our choice of incoming goods.

(3) *Cooked chicken portions.* One stage further and the simplest from a hazard analysis point of view is to buy the chicken ready to eat and then many of the hazards, controls and monitoring become someone else's worry. Your flow diagram then only need include purchase/receipt, storage, plating with the salad and serving (selling), which is much easier to control.

You may of course, for many reasons, wish to cook your own chicken and the point of these examples is not to encourage the use of pre-cooked chicken but merely to demonstrate that it is worth thinking about all your operations with a view to making them as simple as possible. This thinking time could save you time, money and worry whilst the customer gets the same product, in this case chicken salad.

Index

Analysis of food hazards, review of, 52
Assured Food Safety, 69

Baker, flow diagram, 40
Baking, 50
Butcher, flow diagram, 40

Caterers, 15
Catering, flow diagram, 41
Causes of food poisoning, 10
Cereals, 32
Chemical contamination, 35
Chemical hazards, 30
Chicken, 35
Cleaning, 37
Codex Alimentarius, 14, 19
Consumer, 15
Contamination, 43, 64
Contamination, after delivery, 33
Contamination, before delivery, 30
Contamination, chemical, 35
Contamination, physical, 34
Control criteria, 24
Control measures, 33
Control point, 18
Controls, 46, 66, 67
Cooking, 46, 50, 66
Cooling, 45, 50, 66
Corrective action, 26, 51
Corrective action plan, 56
Criteria, 12
Criteria, definition of, 13
Critical control point, 11, 18, 46, 62, 66
Critical control point, definition of, 11
Critical point, 39, 55, 56
Critical points, documentation of, 54

Dairy products, 33
Decision tree, 23, 62
Despatch, 46, 51, 67
Disinfection, 37
Display, 45, 50, 67
Distributors, 15
Documentation, 28, 53
Due diligence, 57

EC Directive, 5, 15, 53, 60
EC Directive, requirements of, 16
EN 29000, 60
End-product testing, 1, 60
European Union, 15

Fillings, addition of, 45, 50, 67
Fish, 33

Flow diagram, 21, 39, 42
Flow diagram, verification of, 22
Food business operators, 16
Food business, definition of, 16
Food handler, 15
Food hygiene, definition of, 15
Food poisoning, 1
Food poisoning, causes of, 10
Food poisoning, chemical, 38
Food-borne infections, 1
Frozen poultry, thawing of, 49
Fruit, 32
Fugu fish, 38

HACCP team, 19
HACCP, and quality, 17, 59
HACCP, documentation of, 53
HACCP, implementation of, 19
HACCP, principles of, 14, 16, 30
HACCP, the law relating to, 15
HACCP, the philosophy, 5
HACCP, the tool, 5
Hazard analysis, 9
Hazard, definition of, 7
Hazard, microbiological, 7
Hazards, 7, 32, 33, 66, 67
Hazards, points where hazards occur, 35
Heating, 36, 50
Herbs, 32
High-risk foods, 37

Ingredients, 32
Irradiation, 37
ISO 9000, 47, 59

Manufacturing, 16
Meat, 32
Meat, frozen storage of, 56
Microbiological contamination, 34
Microbiological hazards, 30, 31
Microorganisms, multiplication of, 37, 44
Microorganisms, survival of, 36
Milk, 32
Milk, heating of, 55
Monitoring, 12, 25, 48, 66, 67
Monitoring system, 25
Monitoring, definition of, 12
Monitoring, documentation of, 56
Moulds, 39
Mycotoxins, 39

Physical contamination, 31, 34
Physical hazards, 30
Pillsbury Company, 4
Poisonous materials, 38
Preparation, 44, 49, 66
Preventive measures, 62
Processing, 16
Processors, 15
Producers, 15
Product description, 20
Public concern, 31
Purchase of raw materials, 43, 48, 66

QA system, 47
Quality Assurance, 17, 59, 60, 61
Quality Control, 60
Quality, definition of, 59

Raw vegetables, 32
Receipt of raw materials, 43, 47, 48, 66

Record keeping, 28
Records, 53
Reheating, 46, 51, 67
Retail food shop, flow diagram, 40
Retailers, 15
Risk, 7
Risk assessment, 9
Risk assessment, definition of, 9
Risk, definition of, 8

Sale, 46, 67
Severity, 9
Severity, definition of, 9
Shellfish, 33
Shellfish poisoning, 39
Small businesses, 29
Spices, 32
Storage, 43, 48, 66, 67
Suppliers, 47

Target level, definition of, 24
Tolerance, definition of, 24
Toppings, addition of, 45, 50, 67

Vegetables, 32
Verification, 13, 26
Verification, definition of, 13

Water, 33

NOTES

NOTES

NOTES